Sacrificing Trees

Lorian Randall

Coco-Randall Publishing
Published by Coco-Randall Publishing
Coco-Randall Publishing, 2074 Robie Street, Suite 2105
Halifax, Nova Scotia, Canada B3K 5L3

Copyright © 2013 Lorian Randall

All rights reserved.

ISBN: 1490949763
ISBN-13: 978-1490949765

No part of this book may be reproduced, scanned, or distributed in any printed or electronic form without permission. Please do not participate in or encourage piracy of copyrighted materials in violation of the author's rights. Purchase only authorized editions.

Dedicated to my three children.
Joseph, Jessica, and Michael
They may be last generation of Geldarts at the "Seashell McCully" on the Fundy Shore.

CONTENTS

	Preface	vii
	Prologue	1
1	Wind and Sea	3
2	My Life Takes Root	5
3	History of this Shore	7
4	The First Geldart	9
5	Clearing the Land	13
6	A Cottage at Shady Nook	17
7	Setting Up Camp	21
8	The Early Years	23
9	The Legacy Begins	27
10	War and Death	31
11	Moving In	35
12	A Dip in the Bay	39
13	Mudflats to Mudpies	43
14	Changing Times	47
15	Water and Roses	51
16	Spring Time	55

17	Stormy Seas	59
18	Alberta Bound	63
19	Reunited	67
20	Who'da Thunk It	71
21	Last Tide	75
	Epilogue	77
	Updates	79
	Appendix	81

PREFACE

Looking back, most of my first memories, and some of my happiest and most meaningful, are from my time spent at the cottage in this tale. I remember shared birthdays with my cousin Lucas, since our birthdays were only a week apart. I had more birthdays at the cottage than at any other location. Growing up, my family relocated often: first across the entire country, then across this province. I cannot point to any one of the houses we lived in as my one true home. No for me and for my siblings as well, there is only one true place we think and feel at home, and that is the cottage.

My mother wrote this book to both rediscover her own roots while being able to ensure her children and any possible grandchildren will know both the Geldart family roots and the history of the cottage. Most importantly, I feel my mother wrote this book to come to terms with the fact that one day, the place she cherishes the most may not exist anymore. Death is the natural end to all things, even the most lived and cherished must come to an end. Until such a time when the cottage is no more, this book will serve as our families "Red Book of Westmarch", ensuring we all will know where we came from.

-Michael Randall, 5th Generation Geldart.

Lorian Randall

PROLOGUE

This is the story of a family cottage in Portapique, Nova Scotia. Portapique is a small community on the Bay of Fundy in the central, eastern part of the province. This story actually takes place on the branch of the bay known as Cobequid Bay.

Having a cottage, or camp, as they are known in parts of Nova Scotia, is an adventure that had become fairly popular in the 20th Century as the Middle Class became more affluent. Many of these started out as family cottages. However, over time, nearly as many have been sold or even worse, allowed to deteriorate. This is not that story.

There are two things about this cottage that make it unique: its longevity and this constant ongoing use by the same family. Also, this is the only cottage that has been in the author's life for over 60 years.

The "Seashell McCully" has survived, not without difficulties, for over 76 years. I think this is amazing in these time of constant change and motion, that we are witnessing in our lives. Granted much of its ability to survive may be linked to the fact that constant change and motion in Portapique or even Nova Scotia in general, is pretty slow, at least compared to the rest of North America. But in truth, the greatest factor that has contributed to its longevity has been the belief, by several generations of a family, in a dream that was started by one man in 1933. This family had faith that the dream was worth holding onto. In other words, they discovered the real value of a family cottage and did all they could to keep that dream alive for future generations.

This story is narrated by an old, white spruce tree which was lost to the tide in October 2008. It was at that time that I had visited the cottage after a huge storm had hit the area with extremely high winds and tides. There lay the tree, on the beach below the bank with its roots ripped out, but still clinging to the soil. Even though I had known for some time this tree would be the next to go, its loss really gave me pause. That tree had been the subject of many photos taken over the last few years. For as long as I can remember, there was always a tree leaning over the bank, tilted towards the tide, ready to be sacrificed. I began to ponder the death of this tree and what stories would it tell if it could talk?

This is that tree's story. So, if there are facts about this tale that are not correct, please, speak to the tree.

Whether or not you know this cottage, I hope you enjoy the story of the "Seashell McCully" and the generations of Geldarts who have lived in her summer shadow. Hopefully, for some readers, the history of this cottage might inspire them to see the real value in cottage life or at the very least, renew or confirm their belief. When we lose out cottages to erosion we are losing part of our heritage.

For me, the author, my joy came from writing about something that has given me many hours of pleasure and from being part of such a legacy.

1. WIND AND SEA

The next tide will probably take me with it and my life here on this water windswept coast will end. I don't mind that it is fall and no one is around to share my final hours. I don't even mind that my death might have been prevented. I only mind that I will never see this new generations of Geldarts as they grow up and grow old.

I hope they will at least miss my presence when they return in the spring. Some even return in the late fall or early winter if the road is passable. They come back to see the winter wonders of Mother Nature and assess her damages.

The wind and the sea have been my constant companions and friends. But, today, as my final tide approaches and the wind howls through my exposed roots and bent branches, I know their friendships have turned mean. The older trees have told me the purpose of the wind and sea is to keep us humble and they have done their job well.

I have known the wind my whole life; feeling its strength since my beginning. Sometimes, in the heat of summer, it offered a cooling breeze and other times, a slap, in fall and winter. When I was younger, I didn't fear the strength of the wind. I was amongst my own family of trees; some of whom past winds caused to look bent and sparse. Their old knar led roots, however, were able to withstand the test of time, clinging to the poor soil that was their foundation.

I have only known the sea fully in the last half of my

life. I say fully because, when I was much younger, I could sense the sea through its sound. To me, the sea was a beast that roared and crashed almost every day. I could hear it at a distance as its sound ebbed and flowed regardless of the seasons. In the early 1930's, the sound of the sea meant little to me.

Today, I know so much more. I know this tide on the Bay of Fundy is the highest in the world. I know its waters come and go up this bay, twice every day, by the pull of the moon.

I know time and tide wait for no man or tree. I know too, the seas has worked with the wind to take much of the land and many other trees and objects that have been in its path. I have known all these things for many years now. Some of these things I have seen for myself and others I have learned from the old trees and from the people who spend their summers on this shore. But what I know best is, I greatly fear this, my final tide, as it races up this rocky shore. This next tide could mean the end of my part in this cottage's legacy on the Bay of Fundy.

2. MY LIFE TAKES ROOT

I am a century old, white spruce tree standing at a height of over forty feet, which as trees go in this area, is pretty much average. My tip is a narrow point resembling a spire on a church steeple. My branches are out at right angles, with the lower ones being dead from living so many years in close proximity to other trees. The seed that gave me life travelled about 500 feet and planted its shallow roots in this less than ideal soil. For most white spruce, the greatest enemies would be forest fires and insects such as the spruce bud worm. Here on the Cobequid Bay, we have other enemies. The ever shrinking land and the increasing level of sea water offer a real threat to those of us blessed enough to over look the water in the bay.

When I began the first thirty years of my life at the turn of the 20th century, my home in Portapique, on the shores of the Bay of Fundy, looked quite different. I was not born into this small grove of trees that I stand amongst today. I was actually part of a forest of trees, including spruce, fir, pine and hemlock,, that covered the shores of the Portapique River right out to the main road that runs between Truro, to the east, and Parrsboro, to the west. The old trees, some in my family are almost 200 years, have helped me to adapt to life on this beautiful and rugged shore.

I suppose, looking back, our purpose in life standing here in this grove of trees was to offer our beauty and protection. The beauty we offered was not the stuff of photography, but rather, in the eye of the beholder. I imagine, only in the eyes of those we sheltered or fed, were we considered beautiful; like a friend with a birthmark, the love generated from the friendship makes even the birth mark beautiful. Some of our homeliness and scars of course can be faulted to the wind, but truthfully, I guess like most creatures, some are beautiful and some are not.

3. HISTORY OF THIS SHORE

The French name for the Bay of Fundy Was "Bay of Gennes" meaning "Bay of Twins". As the bay flows off the Gulf of Maine and makes its journey between the shores of Digby, Nova Scotia and New Brunswick, it branches into two separate bays. Chignecto Bay starts close to Advocate Harbour and continues northward towards Amherst, Nova Scotia. Cobequid Bay veers south, with the town of Truro at its head.

The Mi'Kmaq word for Cobequid was "Wakobeitk" meaning "end of waters flow". The Mi'Kmaq, once called "Mic Mac" by European settlers in the 18th century, are an aboriginal tribe that has lived on the shores of the Bay of Fundy for over 10,000 years. The old ones have told me how these native people found food and shelter being surrounded by these rich waters, lands and marshes. They also used this bay, as well as the Subenacadie River, as a means of canoe travel as far down as Dartmouth, Nova Scotia.

The name Portapique, was originally spelled with "u" after the " a" (Portaupique); but its seems over the last few years the "u" has been dropped. The meaning of Portapique comes from the corruption of the Acadian word "Porc-epic", meaning porcupine. It was also named by the Mi"kmaq to mean "mouth of the river".

After the expulsion of the French from Nova Scotia in 1755, Portapique village was mostly settled by Scottish and Irish immigrants. It was part of what was known as "The Londonderry Region".

In the early 1900's a lighthouse stood where Portapique Beach is today. The lighthouse was built to serve the many boats of the thriving fishing industry of that time. As the fishing industry grew, so did the boat building industry. The best was known as "Halls Boat Building shop" in Portapique.

Another main industry in this Londonderry Region was lumbering. It was the lumbering industry that would draw the Geldart family to this area in the 1930's.

4. THE FIRST GELDART

In the first 30 years I spent on this shore, there was little change with the passage of time other than the changing of the seasons. However, in the 1930's, the sea was to play a greater part in my life. Not only could I smell the salt and feel the spray of the water when the wind drove them in my direction; amazingly, after a savage autumn storm, which felled many trees in our grove, I saw the sea for the first time.

I was awed by these waters in the Bay of Fundy travelling up to forty feet when the tides were at their highest. The low tides were just as amazing with the brown mudflats stretched out as far as the eye could see. The site of the red banks across the bay was breath-taking on a warm summer day with the sun glistening off the dome of a silo on the Noel Shore. The colors and textures of the water were much like the weather here in Portapique, ever changing. When the tides and the winds were calm, the water looked as peaceful and as inviting as any lake. When the tides were rough and the wind howled, the water was a beautiful red brown with white caps covering its surface, jumping up like popping corn.

Each day, 100 billion tonnes of seawater flows in and out of the Bay of Fundy. This in and out flow of the tide, takes a total of twelve hours and twenty-five minutes. So high and low tides are occurring at different times each day and sometimes their change brings a change in the weather. It is truly a miracle to watch this constant movement.

The sea seems to take its direction from the seasons as well as the moon. Of the four seasons, I fear the most the late fall and winter. It is then that the winds and seas do the most damage to the land and trees.

My favorite season has always been spring. Everything on this shore feels renewed and hopeful. One spring especially, brought a great change to my life. On a cool spring day in the 1930's, I saw the first Geldart. His name is Manning Geldart. The old ones told me his home is in Debert about 18 kilometers from this shore. He had been in this area working for the Maple Leaf Lumber Company. Manning was very skilled at dam building and running the river drives in the spring. However, it was not lumbering or dam building that was on his mind that spring day. That day, he was dreaming of the cottage he would build here amongst wind, sea and trees.

I was fascinated as I watched this sturdy, plumb man in his midlife, with his woolen pants held up by wide braces and the dark fedora covering his head. He moved so seriously from tree to tree and from river to shoreline. As he strolled around, he remembered on the move he made from New Brunswick to Debert in 1913, when he was first hired by the lumber company. He reflected on the train ride with his wife, Ada, and their eleven children. They had to rent a boxcar to hold all their belongings. He remembered also, the home he built in Debert for his family after they spent a year at the Coal Mine on Debert Mountain, living in a bunk house. He thought of his now grown children, most with families of their own. Except for the loss of their daughter Lizzie, at the age of nine, before they left New Brunswick, he knew their life had been a good one with many blessings. On that spring day he was ready though for a new direction in his life. What he could not know that day, was the great legacy he would leave for the generations of Geldarts who were to come to this shore after him.

I sensed that he was a smart, hardworking man. I also

sensed his eagerness and a gentleness in his heart towards our family of trees. The old ones had said, that even though the men who lumber cut the trees, it was believed they had a great respect for all of nature's beings, including the trees.

Later that spring, Manning returned to my shore with a man named Spurgeon Carr who owned much of the land and trees here in Portapique. He had agreed to sell a 150 by 200 foot piece of land to Manning for the grand sum of $1.00. Together, they moved around as Mr. Geldart had done days earlier, only this time they used a measure stick and marked off the sold property.

As they shook hands and departed, my curiosity grew.

That night, as the moon floated above with a hazy ring encircling it, and the wind gently stirred my branches, I was lost in thought. As I reflected on the day, I realized with great excitement, life here on this shore of the Cobequid Bay was changing.

Lorian Randall

5. CLEARING THE LAND

Over the years, there have been many trees sacrificed on this shore. Many have succumbed to the perils of Mother Nature. Some have been crippled and even toppled from the strength of the high winds during the fall, winter and hurricane seasons. Others have been destroyed by insects and parasites that feed on them. Many others have been cruelly claimed by the high tides and washed out to sea. The ever growing sea has made their loss even greater over the past few years.

On this day, however, trees were coming down and neither the wind nor the sea is claiming them. It was a warm day in 1933, when spring had returned again to this shores of Portapique, Mr. Geldarts youngest son, Currie arrived. I saw right away, he was a much younger version of his father. Like Manning, he too had a fedora on his head, but, because of the warm weather, it was made of straw rather than felt. Currie had truly followed in his father's footsteps and entered the lumber industry. Manning had passed on his lumbering skills to his son. Even now, Currie is thinking how these skills have served him well, as many men his age are jobless and many families are going hungry during the Depression of the "Dirty Thirties". He presently holds a contract for lumbering with the Maple Leaf Lumber Company; the same company that had brought his family to Nova Scotia.

That day, when Currie arrived, he was not alone. He brought with him, two large, dark, beautiful horses. These magnificent animals had been brought by train from

Western Canada by the lumber company, to work in the logging camps hauling logs. I saw for myself that day, their great strength, as they were able to rip a tree stump from its rooted foundation. I sensed as well the great respect he had for these animals.

Unlike the men before him, Currie returned the following morning shortly after dawn and for several mornings in a row and I greatly enjoyed his company and watching him work.

Each day, he brought with him an axe, a few tools and his food and drink for the day ahead. Everything was carefully laid under an old oak tree that had graced our grove. I would see him stop twice a day to eat from his bundle and relax in the shade of the same oak. When the sun's warmth had increased by noon, I would see him add pieces of discarded clothing to the shade pile or he would simply hang them on an outstretched branch of the tree.

The shade of the tree turned out to be the safest site. One day, a gust of wind, coming off an extremely rough sea, shot his straw hat from the branch of that tree. That playful wind, that had cooled him earlier, now sent him running for several minutes down over the marsh, still boggy and wet from the last tide. The hat went nearly to the Portapique River before he was able to retrieve it. When he finally returned, fedora in hand, he removed his now wet, muddy boots and socks. I smiled when I saw him carefully lay them in the shade of the tree.

On another day, when the Sun thought it was July, instead of May, I saw him slowly wipe his brow, as he had done several times, but rather than continue, he gently lowered himself to the shade of the oak. He lay there relaxing, drifting in and out of sleep. I know, like me, he was watching the many shapes of the few clouds above as they met to form creations of their own. I could feel his peace and pleasure. I didn't know then, I would come to see many more Geldarts, over the years, repeat that same scene.

Currie's days were long and hard. He felled several trees and in the end, he had created a clearing for the cottage his father had envisioned.

Lorian Randall

6. A COTTAGE AT SHADY NOOK

Manning Geldart, his sons, Donis and Currie and his sons-in-law, Jack Graham and Sidney McCully, returned in the summer of 1933 to build the Geldart cottage in that part of Portapique known as Shady Nook.

Jack Graham had married Manning's daughter, Beulah, and they had six children, five boys and one girl. Jack worked as a machinist in Debert and their home was next to Manning's. Their son Ronnie would become an annual visitor at the cottage.

Sidney McCully had married another of Manning's daughters, Lillian in 1928. They now had three children, Leonard the oldest was four, Mae was two and their newest daughter Irene had been born in January of that year.

Manning had passed on his lumbering skills to his son-in-law, Sid as well as his own sons. Sid was working for the Maple Leaf Lumber Company. Sid and his wife, Lillian, worked the logging camps together. Lillian worked as the cook and Sid was in charge of the spring river runs. It was long days of exhausting work and for Lillian, it had to be a bit lonely. The job, however, from most reports, was not necessarily a thankless one. The men spent their days working up huge appetites and a good meal was greatly appreciated. The evenings were often full of music and storytelling.

Because the Geldart family was well rooted in lumbering and working the logging camps, setting up quick and easy housekeeping came naturally. "Roughing It" was the norm, more times than not. It was this ability to "rough it" that helped the Geldart family thrive and fully enjoy their summers on the shores of the bay, sharing their cottage with both family and friends from near and far.

All the lumber for building the cottage had to be brought from Debert, from Mr. Geldart's own mill, where it was prepared. In 1933 that was more difficult and time consuming than it would be today. The work of building was slow, tiring and hot; in others words, it was grunt work, but that was pretty typical for those days. These men were used to a hard day's work and there was no complaining.

The cottage was erected on the shore with 100 feet of land in front, to serve as a safe distance from the tides. The marsh and the Portapique River were to the east of the cottage. The road accessing the cottage ran along the western edge of the marsh and travelers were at the mercy of the tide when coming and going.

The front door faced the sea to the south. Standing on the road and facing the front, you might have thought you were waiting for the local train. With a verandah, running on the south and east sides, with its "Y" shaped support beams, the exterior of the cottage resembled an old train station. The eastern verandah was slightly shorter as a door entered into the kitchen , which protruded out from the rectangle.

The outside was covered in cedar shingles. A stove pipe with a round cylinder running perpendicular on top, rose above the kitchen roof. The roof itself was hipped.

Upon entering the front door, to the left, on the west side, was a curtained off area that served as two bedrooms. A ladder ran up to a loft above which also served as a sleeping area. Because of the raised roof, gases from the kerosene lamps were able to rise.

A fireplace and hearth later stood on the north facing wall across from the front door.

The kitchen was a 10 by 10 foot space with a wood

stove for cooking. The bathroom, like most homes in Nova Scotia at that time was located outside, behind the cottage. It would later become known as the "Rose Bowl" and it took on a life of its own with tales that could match that of the cottage.

Lorian Randall

7. SETTING UP CAMP

A cottage for the working class was not as common in Nova Scotia at the turn of the 20th century as it is today. Certainly, the Geldarts history with the lumbering industry enabled Manning to see a possibility where others might have thought only of inconvenience. His background in working the woods also created a chance for him to choose a piece of land that appealed to his senses. But, perhaps this decision went even further back, to his Geldart Ancestry.

Manning's great grandfather had emigrated from Yorkshire, England in the mid 1770's aboard the ship the "Albion" at the young age of nineteen years. Starting a new life in a new land was certainly not for the frail of heart. Somewhere, inside, there had to be a hope and a dream of a better life, as well as a real faith that you could live that dream someday. The Geldart motto was "Not Always Rain – Non Semper Imbres". Of course, how this motto originated has long been forgotten, but for every generation it has held its optimism.

Throughout the 1930's Manning and his large family made good use of their new cottage in Portapique, especially in the warm months of July and August. Manning's wife, Ada passed away shortly after the cottage was built, but not before she and her daughters had made it comfortable and attractive on the inside.

The kitchen was outfitted with a long table and chairs. A white curtain, made by Ada out of lace, hung in the window that faced the east. This allowed the most light to enter its pane of glass as the sun rose in the morning. Shelves and a small cupboard served to hold dishes that had once been used in the logging camps. These could be identified by the narrow bands of green stripes encircling them. They were heavier than regular china so they could withstand the constant wear and tear. The cupboard also held the supply of food items that had to be brought for each visit.

Because the space in the kitchen was small and the number of people eating at one time could be large, each item, whether on view or not, had to be well thought out, in terms of placement and usefulness. Having to feed eleven children, a husband and quite often unexpected guests, Ada was very adept at this process and the kitchen functioned quite well.

The main sitting room held two rockers, a small table with a kerosene lamp and an ashtray stand with a pipe bowl. A calendar from the local general store hung in the kitchen and a needlepoint sampler graced the wall in the main room.

Kerosene lamps also stood on the dresser in the bedrooms and the beds were outfitted with old quilts and woolen blankets, also from the logging camps.

Although this cottage was not swanky, it was welcoming and comfortable and served the needs of the Geldart family for many generations.

8. THE EARLY YEARS

Manning's oldest son, Donis, who helped build the cottage, was an annual visitor when vacation time rolled around each summer. Unlike some of Manning's sons, Donis worked for the railroad rather than the lumber company. He started out as a cook and later he was a train inspector, or, what was known in the 1930's as a "Hot Box checker". It was his responsibility to see that the friction on the train wheels did not become so great that a fire could possibly break out. He was working a special detail in 1939 when King George IV and Queen Elizabeth I of England travelled across Canada by train. Donis and his family lived in New Brunswick in the Moncton area where he worked for Canadian National Railroad. His wife Greta, had been born in Debert. Donis's sister, Helen Geldart, Manning's oldest daughter, had married Greta's brother, William McCully. So although they lived in New Brunswick, they had strong family ties to Debert and their extended family there. Donis and Greta had only one child, a son, John (Jack).

Unlike his sister Helen and her husband William, who were farmers, Donis had a regular vacation every summer and they were able to travel home to Nova Scotia to visit family in Debert, as well as spend time at Portapique. Each summer, for years, Donis, Greta and their son, Jack, travelled to the cottage, stopping first in Debert to visit. Quite often, Ronnie Graham, Jack's first cousin, whose family lived next door to their grandfather, Manning, was recruited to spend time at the cottage, so Jack, an only child, would have someone to play with.

I looked forward to their visit each year and I especially enjoyed the boys and all their activities. After much unloading and toting inside, the boys were off to explore the area and see what changes had been brought about by the winter storms. Next, it was time to gather some firewood, and fill the water buckets from the well. Finally, time was their own, at least until the first meal was ready. The days were spent mostly outside. There were model boats to build and test, clams to be dug and boiled, fish to be caught, driftwood to be gathered for a bonfire, sticks to be whittled and mudflats and water to be explored and played in for hours. When the milk and bread ran low, there was a walk to the store in Portapique, about four miles, round trip.

Donis and his wife Greta continued to visit Portapique on their vacation for many years. When their son, Jack, was overseas during the Second World War, the school teacher who boarded with them would drive them down to the cottage. This girl was Jean (McGorman) and although she had not met their son Jack, one day she would not only meet him but she would later marry him. Jean enjoyed her time spent at the cottage either walking, reading or helping Donis and Greta can salmon. After Donis's death, Greta, who lived until she was 91 years, continued her visits to the cottage with Jack and Jean and their family. Although new bedrooms were added unto the cottage, Greta refused to sleep in one even when she was in her eighties and the family thought she should have her own room. Instead, she preferred to sleep on a day bed in the main room, where the bed she and Donis had shared would have been in the original cottage.

By the end of the 1930's, Manning and Ada's family of eleven children had been growing steadily. All, except their youngest daughter, Pauline, had married and started families of their own. Most of them, other than Donis, lived in the Debert area. Their son, Percy lived in Truro and their daughter Lou also lived in Truro, about 15 kilometers away. One of their daughters, Adelaide, lived in Green Oaks, just outside Truro. While Donis and Greta were at Potapique on vacation, they were quite often visited by parents and siblings. Both the children and adults enjoyed cooling off in the muddy waters . Even the women donned their bathing suits and allowed the brown, rough water to cool them down. Goodies were often brought and shared. The sharing that really was treasured the most was the sharing of memories and stories of days gone by.

Whatever the worries at home had been, they seemed to be forgotten as these men, women and children, enjoyed their time at Shady Nook.

In the year 1937, there was much grief when Manning's wife, Ada, was to leave her earthly home in Debert for her heavenly home at the age of 67. For her children, she left behind the memories of a great wife, mother and community worker.

Lorian Randall

9. THE LEGACY BEGINS

Although it is spring again in Portapique, my favorite season, it is early and the snow has just melted. I look around at the cottage and her surroundings. The tides have been very high and low, as the moon is in its new phase. The beach stretches out forever with its coarse gravel and ragged shoreline. As a result of the constant erosion, the cottage has only about 20 feet of shoreline left out of the original 100 feet. As I gaze upon this Geldart cottage, twenty years after I witnessed her being built, I see her now in the shadow of winter, like you suddenly see one day that your parents are growing old. The cottage, like Manning Geldart, who is now in his eighties, is showing wear and tear from the last twenty years, from all the families who have visited and the harsh weather she has endured.

A visitor would never mistake this cottage for the fashionable summer home of the well-to-do. Many might be turned off by the cold, muddy waters of the tide and the gravel which covers the beaches and the stubby ground cover that really cannot be called grass.

The Geldart cottage and her home on the shores of the Bay of Fundy is not about fashion or beauty or even convenience. Those things have given way over the years to her true values. Her treasures lie in the familiar sights, sounds and smells and the legacy she has given to all those who grace her door.

One wonderful treasure, for those fortunate enough to find it, is the serenity that comes mostly from being secure in familiar things. There is peace found in the haunting cry of the gulls, the slam of a screen door, the crunch of gravel as feet move on the beach, the hum of insects on a hot August day and the lapping of the waves as a new tide approaches.

There is peace in the familiar sight of the red mudflats and backs bent as they dig to find clams; bathing suits and towels strung out to dry after the day's swim; the rough, brown water on a windy day; the grey, tin water buckets sitting on the porch, waiting for the next trip to the well; the heather moving its purple head in the breeze on the marsh and the scraggly balsam, spruce and hemlock that cling to the soil and dot the horizon.

There is peace found in the familiar smell of a boiling pot of clams, the sweet odor of salted driftwood burning in the fireplace, or the delicious aroma of marsh greens and the fish of the day cooking on the wood stove for a much anticipated meal.

Peace and serenity wrap you in their familiarity upon each visit.

Another treasure is the legacy of the people of this cottage. This legacy, which started with Manning, Ada and their children, has grown as the relationships of friends and families have been developed and allowed to thrive here each summer; as the young have explored her wonders and secrets; as memories were formed, shared and departed with visitors as they left to return to their daily lives. Even those who are no longer with us, are remembered and their stories are retold to the next generation.

I have seen many people come to this shore, the Geldart family and their decedents, other families and friends, the young and the old. Each has had their own hopes, dreams, worries and fears, but, for everyone, if they seek it, there is tranquility and a hope that arises as each new tide approaches and each out going tides takes with it the cares of the day.

Although all who come have their own personal lives, and they may be separated by time and distance, the legacy of this cottage and her familiarity are what unites them as Geldarts.

Lorian Randall

10. WAR AND DEATH

By the spring of 1940, the world was at war and some of Manning's grandchildren who were old enough, took up the call of duty. Donis and Greta Geldart's son, Jack, served with the Merchant Navy. Helen and William McCully's son, Carl, served in the Royal Canadian Air Force. Beulah and Jack Graham had three sons old enough to serve; Glenn and James in the Air force and Ronald in the Navy. Their younger brother, Bobby worked as the telegraph boy in Debert during the war years. Percy and Margaret Geldart's son, Donis, as well as Walter and Frances Geldart's son, Roland, both served in the Air Force.

So those of the grandchildren who were away at war were also away from the shores of Portapique. Although they all returned safely home, I missed their yearly visits and I'm sure they too, longingly remembered their visits and wondered if they would see Shady Nook again. I hoped their memories of the cottage helped to fill the lonely hours spent on far away shores.

Debert, which was home to all these grandchildren and their families, other than Jack Geldart, who lived in New Brunswick, showed many signs of the war in Europe. It was a large military base, both army and air force. It continued as a military base for many years after the war. For many Canadian soldiers being shipped out to Europe during the war, Debert, was the last stop before leaving Canada. Those left at home in Debert had a constant reminder of the war's affect and saw many properties and homes taken over by the military.

Lillian(Lil) and her husband Sidney(Sid) McCully, like most people in Debert, lived in the wake of Army and Air Force activities, as their home was surrounded by military property. Sid, no longer worked for The Maple Leaf Lumber Company. He and his colleague Jim Soy had started their own lumber company, McCully and Soy. With the war on and Jim serving overseas, it was difficult for Sid. It was a challenge to find enough men to work and the first few years were really a struggle, but, the company endured and went on to thrive for many years, leaving a legacy of its own.

Sid and Lil's family also continued to grow. Their five children, Leonard, Mae, Irene, Manning and Linda were all born before 1940 and they were soon followed by the birth of Priscilla, Everett, Michael and the youngest, Donna in 1947. Although their children were too young to serve in the war, death did find its way to their doorstep. Their youngest son, Michael, developed a rare form of lung cancer at the age of ten years. Unlike today, medicine was not advanced enough to offer a treatment or a cure. However, Lil and her son Michael, made a trip to the Children's Hospital in London Ontario by train. Lil had told the family if Michael was with her when she returned, it meant there was nothing that could be done to treat his cancer. Several days later, Lil and Michael returned home by train and Michael died at home in Debert in 1954. Michael had a real love for life and a captivating spirit, so I truly missed him in the summer months. I saw the pain his death caused on the faces of his parents and siblings. His name, however, like Lillian's and Manning's was carried on in future generations of the family.

Death, unrelated to war, also found its way to the home of another of Manning's daughters. One of his younger daughters, Adelaide, known best as Addie, had married Fraser Fisher and together they were farming in Green Oaks. They were also raising ten children. Fraser worked with the Truro Police Force and later joined the Canadian National Railroad Police. On a wet day in February of 1947, he was carrying out an undercover investigation at the Truro train yards when some kind of accident occurred and he was killed. The death was very suspicious but an inquiry failed to determine the cause of death. Addie never seemed to recover from her loss and she died in April of 1948 at the young age of forty-five years. These deaths truly touched many lives in the Geldart family. Addie and Fraser had left ten children that needed homes and love and comfort. They were sent to live with their Geldart aunts and uncles. Two of the boys, Art and Jerry came to live with Lillian and Sid. It is hard to even imagine the grief these youngsters must have felt at losing two parents and their home life. I think their parents would be very proud of the strength and courage they showed and the lives they carved out for themselves.

As the McCully family increased in number, Lillian and Sid wished to give the children a chance to be normal, and vacations were spent at the cottage. The cottage and its surroundings at least offered a reprieve from the personal effects of war and their experiences with death. On this shore, memories of better times and a chance to forget the worries of the day, allowed those left behind to have something comforting and familiar in their lives.

Lorian Randall

11. MOVING IN

When spring returned to Portapique in the mid 1940's, spring returned to the whole world. The Second World War had just ended and those left surviving were trying to pick up the pieces of their lives and move ahead. I looked forward to the summers when Lillian and her growing family returned to this shore for a few weeks.

I always heard them before I saw them. They would arrive in a cloud of dust in a big old chevy truck belonging to Lil's husband, Sid. I could hear the gears shifting as the truck made its way down the old road next to the marsh and I would see the puffs of dust rise, as they neared. The truck would be loaded for bear and quite a sight to behold. Objects and heads peeked out as they pulled up beside the cottage and bodies began to pour out. The cab held the driver, Lillian and the littlest ones. The baby, Donna Lou was the newest member of the Geldart clan and she was held in Lil's arms. Supplies and the rest of the children occupied the open back of the truck. Boxes and bags of food, a new mattress for one of the beds, extra blankets and bedding, towels, linens and clothing all sat piled high. Bicycles, kites and fishing poles were strapped to the wooden side rails of the truck.

Although their home, Debert, was only a village, it was very active as a military base. Many people passed through this community during the war years. Here in Portapique, there were no best friends to play with – only siblings, the nearest store was a couple of hours by foot, the food supply was limited, with no refrigerator, and the sleeping quarters were tight. But, none of these concerns registered on their faces as they jumped from the truck and ran to see the tide.

Before the actual fun could begin, the truck had to be unloaded and a place found for everything. Sometimes, there was a bit of arguing over who was sleeping where, with the first choice often being a site outside on the verandah or in the large, canvas, bottomless tent that would be erected. After a couple of hours it was all sorted out and the cottage was aired and things put in their place. It was at that point the truck and driver would leave and Lillian and the children were on their own until the first visitors arrived.

Some would say Lillian was very brave. She was that, and she was also very devoted to her family and their well being. She truly felt a special connection with nature and her Lord here in Portapique. At home in Debert, she loved to spend time gardening, especially growing roses. Here at Shady Nook, there were only children to tend, but, like her father Manning, she had great love and respect for the trees. Even when she was in her later years of life, she visited Portapique each June and always encouraged the planting of new trees to replace those lost over the years and to help protect the cottage and shoreline.

Sometimes Lillian's father Manning, would come down for a visit and could be seen sitting on the front verandah smoking his pipe, as he enjoyed all the activities around him.

Sometimes, her sisters and brothers and their families would arrive. These visits were always well appreciated as it sometimes meant special treats when the food supply might be running low. The McCully children especially enjoyed their visits from Uncle Percy, when he would bring his home made root-beer.

Food was not as difficult to come by as you might think, but, you had to be opened minded and resourceful. When the fishing boats were returning home up the bay after a day of fishing, Lil might send one of the children down to the water to purchase the catch of the day. This fresh fish, either, shad, salmon or bass, along with a feed of marsh greens made for an easy and nutritional supper. In late June and July there were strawberries to pick and in August there were blueberries. Of course, a favorite food for many was the clams which were there for the digging when the tide was low. Perishables like milk and butter became very precious as they were hard to keep fresh without a fridge, It helped a great deal when the ice truck eventually made its way down to Shady Nook. Originally, a hole was dug next to the well and a barrel or lader, cement crocks were inserted and covered to keep the perishables cool and help increase their life span.

For the McCully children, the days were usually spent outside at various activities. Inside activities might include reading a book or playing a game such as "checkers" or "snakes and ladders". A jigsaw puzzle was also a great way to pass the time and all these activities are still popular as the cottage has no permanent television. A wise Lillian did not allow her children to play cards or read comic books at home, so many hours were devoted to these activities while at the cottage. There was little down time and when it came, it was usually needed. Lillian was famous for making everyone have an afternoon nap whether or not they needed it. If you couldn't sleep, at least you had to lie quiet. This often caused much distress for the children, who had things they wanted to be doing and saw resting as a waste of valuable time. This tradition carried on with her grandchildren as well and it was just as unpopular with them. I guess only the adults saw the true advantage of siesta time. There was one upside though. The children, at least those on the ball, could argue they should be able to stay up late because they had a rest earlier. Sometimes, this reasoning worked on the adults, but, usually, as they learned in time, not on Lillian.

This ritual of moving in continued for many years in

the McCully family as the older children grew and started their own families and rituals.

For one family in the 1980's, moving in took on a special meaning. Lillian's granddaughter, Lynn and her husband Robin and their four young children spent the summer of 1986 living in the cottage while their new home in Crowes Mills was being built. They were relocating from Parrsboro and had sold their home there, so they needed shelter while the new home was under construction. I am sure it was less than ideal, as it was one of the rainiest summers on record, The children, however, had a great summer. They built rafts, climbed trees, swam and explored the area, undaunted by the weather. It seems weather has more importance for adults than children. I delighted in watching them as they played and learned to swim and became the next generation of Geldarts.

For some, moving in was just a stopover or done on a whim. In this case all you needed was a key to enter and every Geldart knew where that was kept. Such was the experience for Donis's son, Jack when he and his new wife, Jean, stopped into the cottage in 1952, on their way home to New Brunswick from their honeymoon. Too weary to continue their journey that night, they found a sleeping bag up in the loft and spent a restful night in Portapique.

Over the years, many others have stopped in for short visits. Some come to spend the day, to fish the tide or maybe to share a meal. Whether a trip to the cottage involved a major move or simply finding the key and entering for a while, I always enjoyed the company, no matter how long they stayed.

12. A DIP IN THE BAY

For those who have never swam in the Bay of Fundy, the first time can be a bit of a challenge. The muddy waters, the rough waves and the cold temperatures of the water might be a deterrent for the faint of heart. But, for those raised on this shore and summer visitors, these are just a normal part of enjoying a dip in the water. Whether they were young or old, the Geldarts made the most of the full tide, especially in July and August.

Morning dips were even on the agenda for some of the more dedicated swimmers, if there was water in the bay. Some swimmers had a preference for only the incoming tide, letting the current carry them eastward, towards the head of the bay. Those who stayed in until the tide had turned were pulled westward, toward the mouth of the bay. Whichever way the tide was going there was constant motion and this was not to be taken lightly. The higher the waves, the warmer the water and the greater the squeals and shouts of enjoyment.

It was entertaining to watch each of the Geldart swimmers and see their individual rituals develop. For Sid and Lil's boys, an incoming tide was the signal to head to the woods so the swim time could be avoided. For the more enthusiastic swimmer, it was the "run and jump right in" method and for the squeamish, the "tip your toes and scream" method. The "old inching your way in" method with constant coaxing from others has remained the favorite over the years. Whatever their method of choice, once immerged, I could see the pleasure on their faces and hear their shouts of enjoyment as the current and waves tossed them about.

Some swam far out while others held to the safety of the shoreline. From time to time, as a new generation of Geldarts came along, the youngest might be plopped at the water's edge and they were often toppled over or splashed by the incoming waves. Although, initiation into the Bay of Fundy often resulted in tears, as the young ones grew so did their joy in the water. Sometimes, youngsters, not ready for independent swimming, could hitch a ride on mom or dad's back.

For a lot of swimmers, getting out of the water could be as much a challenge as getting into the water. Some days there were those large moose flies hovering over the water, just waiting for a taste of flesh bathed in warm salty water. Their bites must have really stung because the cries could be heard at a great distance. Then, there was the gravel beach. Getting down the beach with dry, bare feet was easy for the seasoned veteran but, going back up with wet, bare feet was not a pleasant experience. If you lingered too long after the tide had turned your walk was increased in length considerably. Wearing footwear, even just to the water's edge was great, as long as you were careful how close they were left to the incoming tide. The same was true of that towel they dropped on the way in. Some things just never change and I have seen that moving tide claim many shoes and towels over these past seventy years. However, when the weather was hot and the sun beamed down in Shady Nook, all the sacrifices made for a refreshing dip in the bay were well worth it.

With swimming goes swimwear. The issue of bathing suit availability could sometimes present a problem and more than once a still very wet bathing suit had to be put on to get in that second swim of the day. Most of the Geldarts, I'm sure, have memories of how sticky and tight, not to mention cold, a wet bathing suit feels when there is no alternative. In the most desperate times, an ill fitting suit had to be donned when swimming was the order of the day and you didn't want to be left out. Many laughs were shared when the swimmer strutted from the cottage wearing a suit that was twenty years too old, the wrong size or worn by the wrong sex. It was all in good fun and the desire for a good swim was very powerful. Whatever the color of the suit was when it entered the water, chances are it was reddish brown when the swimmer emerged from the tide and there might even be a little scoop of gravel in the bottoms.

Over all these many years that I have watched the Geldarts swimming, young and old, Priscilla has been the most faithful swimmer. She is Manning's granddaughter and has been swimming here for over sixty years. I remember her first introduction to the water in the late 1940's and I knew then, this would be a long love affair with the bay. Of course, she had older siblings who served as role models, well, at least her older sisters did. When Priscilla was four years old, she had to have a tumor in one of her eyes removed as it was cancerous. The surgery involved the removal of the eye. Although glass eyes were available, they were costly for the family (about $12.00 at that time); so it was necessary for Priscilla to use caution during certain activities. When she went swimming, she was not to submerge her head under water. This sounds simple enough, but, the roughness of the tide made that difficult to prevent. One hot afternoon while swimming, the inevitable happened: the eye came out and was lost in the sea. Lillian, Priscilla's mother, was very resourceful and when the tide was flat, all the siblings helped Priscilla comb the gravelly beach in search of the missing eye. This was quite a daunting task but the eye was recovered and a sense of victory ensued, especially for Lillian. Lil, herself, never actually swam in the bay after she had children. She was, however, a diligent lifeguard whether from the shore or with her feet in the edge of the water.

Priscilla still remains very faithful to this shore and her

daily dip in the bay is still one of her favorite pastimes. I still look forward to her swims, quite often with a grandchild in tow.

13. MUDFLATS TO MUDPIES

A day at the beach in Portapique may not compare to other shores with their sand dunes, sand beaches and warmer water, but, if you were raised on this shore, or just willing to spend the day, there are pleasures to be found here on this gravelly, sloped beach with its ever changing tide.

When the tide is low, the curious explorer can find things exposed that have previously been hidden.

For some, it might be a leisurely stroll watching the mudflats spread out further and further. At times, it seems as though you might be able reach the Noel Shore directly across just by walking from mudflat to mudflat. While this may be tempting, most Geldarts know, or, learned firsthand, the tides can be tricky and can slip quietly in behind you once they have turned, leaving you stranded on a mudflat that is quickly loosing it's ground.

Amongst those who enjoy beach combing are the collectors. Not everyone starts out a collector, but it can become addictive, as the Geldart family and others have learned ,and, there are definitely no age barriers. Visitors, especially, delight in rock collecting, mainly because the selection on this gravel beach is endless and ever changing. The shores of the Bay of Fundy are home to minerals of all sorts with two of the most common being Agate and Amethyst. Agate is the provincial gem and amethyst is a crystal known for its healing abilities. The average Portapique rock hound, however, is happy in finding the many shapes, sizes and colors of the ordinary stones. Many of these are lovingly carried back to the cottage to be shared with all and sometimes even painted by the more creative collectors.

For others, the search is all about seashells and beach glass. At least these are a little lighter for transport back to the cottage. Once again, there is the reveal and display set up for everyone's enjoyment. Unfortunately, as the years have passed, sometimes these treasures are lost among the garbage that washes up with the tide and the beach glass maybe nothing more than someone's broken beer bottle. Those passionate about the beach have become garbage collectors as well. But, it is all just part of the journey for those hidden treasures.

Collecting driftwood is also a great pastime and there are certainly connoisseurs. Collecting driftwood is not done solely for stocking up on firewood or getting just the right stick to whittle. To the connoisseur, this activity involves finding pieces of wood that make the collector see something else entirely. Like most treasures, these are proudly displayed and sometimes even mounted and varnished for longevity.

For many years, Manning's two granddaughters, Mae and Priscilla, were both very devoted collectors. It can take a keen eye to see visions in a piece of driftwood and not all are as blessed or as dedicated as those two girls. Their joy and passion in their finds continued into their adult life and they inspired their children and grandchildren, some of whom are great collectors in their own right.

Another fascinating site for discoveries is the mudflats themselves. Hours are often spent with backs bent looking for treasures on, or buried in the warm, soft mud and clay. Sometimes the containers brought back held both living and dead creatures that had been harvested. These too, were proudly shown off and safely guarded, at least until the next low tide. Collecting the gray potter's clay that lay buried in the mud banks was a special activity, one that Mae, Lillian's oldest daughter, was fond of doing. Once back at the cottage, the clay could be molded into objects and decorated with stones and shells and put in the sun to dry.

But, for the young or even the young at heart, mud sliding has been a favorite outing. Where the marsh and the mudflats meet, there are banks up to six or seven feet in height composed of the most delicious looking, sun baked, soft mud. Your bottom is your sled. Again, hours can be spent coating your body in mud.

By the end of the afternoon, I would see the redness on their backs and arms from the heat of the sun, even under several layers of mud. Although the first layer of mud is light brown with the darker layers underneath lightening as they become older, it is sometimes the first layer that is difficult to remove, as it is totally sun baked and dried.

Washing up after a trip to the flats is certainly not a pleasant experience and can present a real challenge, especially before there was running water at the cottage. A light rinse is sometimes done at the water's edge but with the tide out there is only the icy cold, well water straight out of the hose. For many, this brings howls of shock and shivers, especially on that newly burnt skin. Despite the torture of clean up, talk soon resumes about the next trip to the mudflats.

Lorian Randall

14. CHANGING TIMES

By the 1950's, the outside world had undergone many changes in terms of politics, technology and attitudes. The Cold War was on but so was the baby boom. With the world wars behind them, people felt a renewed hope for a better life. There was a generous supply of goods compared to the forties and although families were a bit smaller, sometimes in number, the beginning of the age of the nuclear family had begun.

Changes on the political scene were evident. King George IV had died and in 1953, his daughter, Queen Elizabeth II took the throne. Ike was in the White House and Louis St. Laurent was the Liberal, French, Prime Minister of Canada. Russia and Communism were the new enemies and the latest in warfare was the fall-out shelter. The cost of mailing a letter was a hefty three cents.

New advances in technology were evident everywhere, from washing machines to hairdryers and all these advances helped to make domestic life easier. Television was making its way into many Nova Scotia homes. Gunsmoke had begun its long running series in the 1950's and people anxiously tuned in each week to see Matt, Chester, Kitty and Doc as they tried to keep law and order in Dodge City. Shows like the Lone Ranger and Roy Rodgers were entertaining the young and old alike. The success of the movie, "The Egg and I" brought to life , on the big screen, a series of comical movies called "Ma and Pa Kettle". The new technology reflected a more relaxed way of life.

Changes in attitude were also evident. In 1956, Rosa Parks, a black working woman from Montgomery, Alabama, refused to ride any longer in the back of the bus, creating the path for the end of segregation.

The music also moved into a new direction with the birth of "Rock and Roll", although attitudes about this new music were slower to change with the older generation.

The kids of the day, who were bound for no good, if they listened to that new music, wore bobby socks and saddle shoes. The girls wore poodle or flared skirts, pedal pushers and ponytails. The boys wore blue jeans and ducktails. Eyeglasses, if you were cool, were cats- eyes glasses, hip enough to make you wish you wore glasses. Scarves for women were all the rage. As this new generation of adolescents moved to the music and spent free time making their hips move the hula-hoop, many older heads were shaking.

During the 1950's, the cottage at Portapique was also changing.

In 1955, the shoreline had eroded so much the cottage itself had to be moved back from the brink to save it from an untimely death. Two new bedrooms were added on the north wall and a new fireplace was built on the west wall by Albert Vance of Masstown, Nova Scotia. This opened up the main room considerably although the loft was now gone. A new road was built by Lillian's husband, Sid, using his own heavy equipment and crew. Finally, the cottage could be accessed at any time, regardless of the tide. This new road ran parallel to the original road but on higher ground, just to the west of the marsh.

Manning had purchased, for the sum of $50.00, the rest of the shoreline out to the highway running between Portapique and Five Houses. This land was later deeded to his grandson and name sake, Manning McCully and the cottage was deeded to his daughter, Lillian.

There was another cottage built by Manning's son,

Currie and his wife Margaret on the same lot as the original cottage. It later became known as the "Driftwood Saga".

One of the most exciting changes occurred on the Portapique Beach Road, across the river from the cottage. A new dance hall had been built there and a Saturday night trip over to the dance hall to hear the "Barley Boys" play, was a great outing for the adults.

The Geldart and McCully families were still constant visitors to the cottage. Travel was more available and convenient throughout the 1950's, so more family members spent their vacations in Portapique. During the years of the fifties the cottage became known as "The Seashell McCully". Sid and Lil's oldest three children, Leonard, Mae and Irene were married and had started families of their own, who also enjoyed their cottage visits. Manning's number of great grandchildren continued to grow throughout the 50's, 60's and 70's. Each new member of the extended Geldart family had a chance to experience the legacy of that cottage, built so many years before. Quite often, those who lived the closest got to frequent it the most.

Some things, however, just never changed in Portapique.

The clams had to be dug at low tide, the water buckets still had to be filled from the well and the wood for the stove and the new fireplace still had to be gathered. There were boats to try out, fish to be caught, swims to cool down, exploring on the beach to do, walks to the store in Portapique and games to be played.

Card games were enjoyed by all and were a great evening's entertainment. One Saturday evening in July, when Lil's daughter, Mae, her husband Laurie, as well as other family members were relaxing with a night of playing cards, a car pulled in and a knock came to the screen door. A gentleman was looking for Portapique Beach Road, which ran on the other side of the river. He had heard about the dances and was anxious to attend. Laurie told him to go back out the road to the highway, turn right and he would soon see the road. The gentleman left but returned again in a few minutes and could not believe he was back at the same screen door with the same people playing cards. He had followed the directions, but, he hadn't gone far enough down the highway and ended up taking the old marsh road into the cottage. His response, when he saw where he had landed was, "How many damn roads are there into this cottage?" Everyone had a good laugh and Laurie retold this story for many years to come.

Another constant at Portapique was the sight of Manning Geldart, sometimes with his friend Eugene Fulton, sitting on the front verandah smoking their pipes and maybe whittling. They so looked forward to any visitors and any supplies they brought, especially if that included sweets and rum, or any spirits.

Manning, who was now in his eighties, lived in Debert with his son, Currie. He had a great love of baseball and often arrived at the local game in his old black Studebaker, still wearing his fedora and smoking his pipe. He was a loyal fan of the Debert team and if they started losing, Manning would leave, grumbling that the empires (umpires) were no good! More than once, the baseball team was invited to Portapique for a picnic and a big feed of clams cooked on a bonfire.

15. WATER AND ROSES

In the first half of the 1900's, outhouses were as much a part of life in Nova Scotia as were clotheslines and woodpiles. Daily and nightly trips to visit the "john" were a standard occurrence and simply a fact of life, especially in the rural areas. Things were no different in Portapique.

The outhouse stood behind the cottage at a suitable distance. It was a small wooden structure and was very basic, just a one-seater. Its roof sloped to the back with a small rectangular window above the seat, simply covered with a piece of screening, at least in the good times.

When the Geldarts arrived in the spring time for the yearly opening and cleaning, low on the list of favorite chores, was cleaning the outhouse. Although it was a smelly job, it was probably the quickest. A sweep out with the broom to get rid of any debris, a bag of lime down the hole, the latest copy of the Eaton's Catalogue and it was business as usual. Longer term maintenance required the path to be kept safe and clear, especially for those nightly visits and a coat of paint outside every few years.

Over the years, the outhouse at Portapique took on its own identity, separate yet similar to that of the cottage. Firstly, like the cottage, it acquired its own name, "The Rosebowl" and a sign above the door announced it to all visitors. Now, roses certainly could have grown in the fertile soil and by times their fragrance would have been a welcome relief, especially on a hot summer day, but, the only roses left in the bowl were pretty unpleasant.

In 1955, the Rosebowl had to be relocated when the

shoreline had eroded so much that the cottage itself had to be moved back. A bathroom so close to your back door might be convenient but not all that practical. Unlike the well, which had to be moved when salt found its way into the water supply, the Rosebowl was happy to serve wherever it stood, and serve it did, for many decades.

Unlike the outhouse at home, Portapique's outhouse was of the fair weather variety, so structure was not as important, as far as warmth. There were gaps and knots everywhere, adding to its charm, but, cutting down on the occupant's privacy. Often these holes were stuffed with whatever was available, whether it was toilet paper, chewing gum or old rags. Over the years, the walls themselves began to talk and tell stories with many messages and verses left by the more creative Geldarts, especially those of the adolescent persuasion. Old pennants and license plates also found a home on these walls.

Visitors to the Rosebowl could wile away their time searching the pages of the catalogue or any reading material fit for the outhouse, keeping in mind that when toilet paper ran out, any paper was fair game.

A small hook and eye closure on the inside ensured no one would enter. A rectangular block of wood with a nail through its center was attached to the outside doorframe. If the block was sideways across the door, it was free to open and enter. If the block was straight up and down, that was the signal that it was occupied. However, as it often turned out, someone else entering was not always the concern of the occupant.

Aside from its practical use, the Rosebowl developed two other useful functions, especially for the young at heart. First of all, the sign that it was in use, was often a signal to those outside that playtime could begin. By simply turning the block sideways across the door, the unsuspecting user was trapped inside. This was not a great place to be on a short term basis at the best of times, with insects and odor as companions. If it was night time, the urgency to escape was even greater. It was fascinating to see how family members reacted and who showed up for the rescue. More than once, I saw that old, wooden block ripped through the nail holding it in place, as the occupant gained their freedom with force.

The block and nail had to be replaced many times over the years, but it was not always the result of trickery. I remember one of Manning's younger grandchildren slipping away from the others to retire to the Rosebowl. A few minutes passed and my attention was drawn that way again as I heard desperate cries for help. I realized the wooden block had become loose and slid down, covering the door enough to trap the child inside. It was several more minutes before the wailing could be heard above the wind and the waves. Help did eventually arrive and although no one was physically injured, a partner was required for the child's next visit to the Rosebowl.

The second function was the water fights which quite often involved the outhouse. The Rosebowl, as it turned out, was a great, "hit and run" site. A bucket of cold water through the window in the back was sure to soak the user and bring gales of laughter from the on lookers as the victim emerged at lightning speed. The goal of the water fight was to catch your opponent unaware and such a high stake game was enjoyed more by the adolescents and adults, who were young at heart. First of all, the water was icy cold as it had to be carried from the well. The payback could be serious, especially if you had to take a full bucket of water in the face or down over the head, when you unthinkingly turned a corner, stepped out the front door or went to the outhouse. Having a good hiding place was critical and the Rosebowl was not the answer.

By the end of the 1970's, life styles in Nova Scotia had changed. People started eating out doors in the summer and using the bathroom indoors. Indoor plumbing eventually made its way to the Seashell McCully, but for many years following, the outhouse still stood her ground for usefulness. It served for emergency uses when the indoor bathroom was busy and for some of the family, old habits just die hard. When you were swimming and nature struck, no one cared if you dripped water all over the outhouse. The techniques of good spying could be honed trying to peek through the knot holes without giggling.

Throughout its history, the Rosebowl always remained a one-seater, although it was once modernized with a fancy plastic seat, like those indoors.

Finally, in the mid 1970's, the Rosebowl was literally laid to rest when the cottage once again had to be moved and indoor plumbing was installed. I still miss all the visitors and activity that thrived around her. At least the water fights still continued.

16. SPRING TIME

Each new spring, I looked forward to the arrival of those Geldarts who came to open and clean the cottage. Of course, over the cottage's many years, the Geldart cleaning crew has changed many times. In the 30's when the cottage was new, Manning and his wife Ada looked after this chore, but, after Ada's death in 1937, Manning and his children carried on this yearly ritual. In the 40's, Lillian, her own family and her many brothers and sisters picked up this duty and by the 1950's and 60's, this job was generally carried out by Lil and her now grown children. Her married daughters, Mae and Irene and their respective husbands, Laurie Matthews and Aubrey Gratto, were often here to lend a hand. Some years, many family and friends joined in making the task one that could be completed in a day. The McCully girls often wielded paint brushes, sprucing up furniture, walls, signs and anything else that was fair game for painting. Sid and the McCully boys, as well as Art and Jerry Fisher, looked after any repairs to the building or road. By the 1970,s and 80's, the McCully families were helping Lil and Sid more and more. By the 1990's, Lil, who was in her eighties, relied on her family to take over spring cleaning. Because many of her children had married and left Debert, this chore fell mostly to her daughters Irene and Priscilla. Priscilla now had her own cottage in Shady Nook , but, she and her husband, Eric faithfully tended to the up-keep of the "Seashell McCully". Irene and Aubrey also had a large role to play. Lillian had deeded the cottage to Irene upon her death. Donna and her husband also loved to spend time at Portapique and donated a lot of their time to painting. Sid and Lil's other children would join in when possible. Although Lil's cleaning days were behind her, she often had big and little projects she wanted completed. She always looked forward to her weekly visit in June and I always enjoyed her time spent here. Until her death in 2003, she held the cottage very dear to her heart and I held Lil very dear in mine. She was a real champion of the trees and always encouraged family members to plant and care for new trees around the cottage. Like her flower gardens in Debert, she tended us trees with the utmost care and concern. Lil was also the Geldart who worked the hardest at keeping the legacy of the cottage alive for future generations. This was never done as a chore, but, as something she loved and believed in. She had truly found the real treasures here and she wanted each new generation to have the chance to experience life on this shore in Portapique. Because of her dedication, the joy Lil found here, has been passed to her grandchildren and great grandchildren and hopefully will continue for many years to come, even though I, like Lil, may not be here to witness those times. Lillian passed away in January of 2003 at the age of 92.

Lillian, from the time she was young, had her loves in order; her Lord, her parents, her husband, her family and her community. This proved a winning recipe for a good life. Over the years, Lil stayed very faithful to her loves and their order, and because of that, her loves were very faithful to her. Her husband, Sid, was certainly a great example of how a love for your wife and children, brought many blessings to a family. A quiet, dignified man, Sid supported Lil in all her loves. It was no different when she developed a strong love for the cottage in Portapique. Because of Sid's devotion and hard work, the cottage was able to survive and flourish.

Although it has been five years since she last visited

this shore, sometimes it is like time stands still. I often recall her sitting on the verandah, as her father before her had done. I can see her shock of snow white hair, her cane leaning against the rocker, and her warm smile as she greets family and friends who have come to visit.

When it came to spring clean up, all helping hands were appreciated and the faces changed from year to year but the rituals were pretty much the same. There were mattresses to be aired or replaced, dishes to wash, floors to scrub and insects, mice and their debris to remove. Some years there was new furniture, new curtains, new pictures and new puzzles and games to be attempted.

One spring in the 1950's, Donis Geldart hauled a small trailer behind his car and brought from New Brunswick, some furniture and fixtures for the cottage. These included an old tin cupboard that had been used as a display case in a general store. Two metal chairs that were often taken outside to the verandah and several oil lanterns that were attached to the walls. These lanterns had previously been used as a means of light on the railroad cars. His wife, Greta, brought a framed poem and picture. Donis's daughter-in-law, Jean, remembers that picture hanging in the room she rented from Donis and Greta when she taught school in their community. I believe the poem sums up the mood of this cottage very well.

"Just hang your hat on any old peg
Your coat on any old hook,
Seat yourself in any old chair
And read from any old book;
Sleep and eat whenever you please
And through the whole house roam,
Consider yourself as one of the folks
And make yourself at home."

Most years, this great clean up took place in May, if weather permitted, on the Victoria Day weekend and I hope it will continue for many years to come.

Lorian Randall

17. STORMY SEAS

The legacy of the Geldart cottage continued for 40 some years, until the year 1976. Looking around, Shady Nook, I was sure the life of this cottage would be over.

Although spring had returned to Portapique, I was not delighting in the sun's warmth, the new growth of the vegetation or even the anticipation of the return of the Geldart families.

A late January storm had beat its way up the coast of the Cobequid Bay leaving in its wake, not only the remnants of such weather, but more tragically, little hope for summer and watching all the activities that have been my joy over these last many years.

Because the weather in January had been quite mild, known here as a January thaw, there was only a scattering of ice in the bay. Without ice, there was no protection for the land when high tides and gale winds hit in that January storm. It lasted for two days and nights.

The sight that lay before me was heart breaking. The cottage, which had once resembled a little train station, now resembled a long abandoned shack. It's proximity to the water's edge was scary, making my own demise more looming. She was perched on the lip of the bank, her verandah lay in pieces, scattered over her yard and some of her roof was missing. The yard was littered with debris, including some trees that had been uprooted.

The bank itself had been ripped jagged by the stormy seas. Great clods of dirt and sod hung over the bank below the cottage. Driftwood from other shores had been left helter skelter on the beach and land.

In my dismay, a hope for another summer with the Geldarts and their families was dashed. Spring had returned to Shady Nook, but would the legacy continue?

But, the Geldarts did return. Manning's daughter, Lillian and her husband, Sid, came early that spring to access the winter's damage. The shock and pain on their faces when they beheld the ghastly sight before them, made my heart ache. I was thinking how unfair this situation was and that it would be best to let this beaten, broken cottage go out to sea in the next stormy tide. I know, the same thought must have occurred to Lil and Sid, but their actions proved different. What nature had done might not be fair, but, that was not the issue now. The family rallied and it was decided: a second move was in order. That old motto "Not Always Rain" surfaced and it proved as optimistic as it had in the 1930's, when the dream began and in the 1950's, when the cottage was first moved.

Manning Geldart had passed away in 1960 at the age of 93 and I am glad he never lived to see that devastating sight of his beloved cottage. I know he would have been proud of these McCullys as they undertook the business at hand. Not only was the cottage moved back, but, it also received a new verandah and a new roof as well as indoor plumbing. Everyone chipped in where they could.

Mae's husband, Laurie, had gone down to do some carpentry work and he had taken one of his nephews, Noel, Irene's youngest boy, with him to help out. They worked hard in the daytime and lived on hamburgers and hotdogs. In the evening Noel and Laurie, who was noted for teaching all the kids to play cribbage, had several rounds of crib. When Irene dropped in for a visit after they had been there a couple of days, she asked Noel how he was making out. "Oh, mom," he said, "We are living like kings".

In the end, the cottage looked better than ever.

Sacrificing Trees

Although I was no longer worried about the safety of the Seashell McCully, my own safety crossed my mind from time to time. I still had a good thirty feet between the bank and my roots, but from the past forty years on this shore, I have learned the tide can be greedy when it comes to the land and sacrificing a tree is a common event.

Lorian Randall

18. ALBERTA BOUND

The summers of the 1970's were, as usual, very busy ones here in Shady Nook. Although Manning Geldart passed away, his daughter, Lillian, kept his wish alive, that the "Seashell McCully" be a family cottage shared by all, for the remainder of her life.

Mae and her husband, Laurie and their girls, Lorian and Lynn, were frequent visitors to this shore. By the mid 70's, Lorian was teaching school in Hants County and Lynn had married Robin MacLachlan, from Great village. Mae and Laurie both enjoyed the life style the cottage had to offer and because Mae now lived in Kentville, it was a great way to visit with family and friends from Debert. Mae was an educator and also an artist in her free time. When she came to Portapique, she often brought her paints with her and what she painted on did not really matter. Whether she used old pieces of wood or canvas, she let her creativity be her guide. Some of her creations still hang inside and outside the cottage today and if you are curious and look behind one of them, you might discover another older painting. Mae took such pleasure in her creative abilities and she was always more than willing to share her activities with her siblings, children and others. Her creative eye allowed her to see materials around her that could be used for art. She would gather the gray clay buried in the mudflats and create objects to be dried and painted. Seashells and drift wood were painted and shaped into various designs. Even wood could be whittled.

Mae also enjoyed the time spent with her Uncle Currie,

who had built the "Driftwood Saga", next to the "Seashell McCully". Currie called her "Maisey" and they loved to tease each other. It was entertaining to see who would get the last laugh. One summer in the 70's, streaking was making headlines everywhere and it became quite the topic of conversation. Currie, who was getting quite a chuckle out of all the streaking occurrences, had spent the summer tending to some pole beans he had planted in his yard at the cottage. That year, Mae painted, on an old piece of plywood, a mural of Currie, streaking through his pole beans. Currie was so tickled by the painting and hung it on the outside of his cottage. In 1978, Pope John Paul I, died very suddenly, thirty some days after his election, so a second election was held the fall of that year. Currie loved to tease Mae and tell her what color the smoke from her chimney should be, even though they were both Baptist. Currie was very fond of playing horse shoes and cribbage and he and Laurie spent many hours doing both. There was a clam factory in Five Islands, about half an hour drive away, and Curry often arrived with a big feed of clams to share.

In 1979, Mae's oldest daughter, Lorian accepted a teaching position in northern Alberta and her last night in Nova Scotia was spent with Mae and Laurie In Portapique. She was driving across Canada for the first time by herself in her little yellow Honda, which was loaded for bear. While Lorian was excited about this new adventure, she was also very nervous. Although her parents were pleased with her decision, they too were worried about this long journey by herself. They had all shared so many happy memories at this cottage and its sense of peace and goodness helped calm some of their worries on that last evening. It was assumed, as they watched her drive off, that they would see her at Christmas time and the two years she planned to be away would go by very quickly. However, as is often the case in our lives, things don't always go as planned. Lorian met her husband Rick while teaching in Alberta and those two years became ten.

In 1986, Lorian and Rick were expecting their third child. This baby, Michael, unlike their other children, Joseph and Jessica, would be born in July, when Mae was on vacation. She would be able to fly to Lethbridge at the time and there was much anticipation around this birth. But, life suddenly took a cruel turn. On Easter Sunday of that year, Mae was lost to us all in a fatal car accident near Windsor, Nova Scotia. Of course this family had experienced death before, but the suddenness of this death seemed to shake this family to its very core. Not only did Mae never get to see the birth of her grandson, she was never to visit this shore again. Everyone felt the gripping pain of this death for a long time.

Here in Portapique, the signs of her passion for art, although a bit worn and faded, still live on. I, like everyone miss her dearly, but I feel her presence in her children and grand children.

In 1989, three years after the death of her mother, Lorian and her family moved back to Nova Scotia. Although her children don't remember their grandmother, Mae, they have been able to understand who she was because of the time she spent on this shore.

By the fall of 2003, another of Lillian's daughters, Irene had deeded the cottage over to Mae's daughter Lorian and it has been under her care ever since with lots of help from family members.

Lorian Randall

19. REUNITED

Over the past 76 years, the Geldart Family has increased, a great deal in its size, as most families of that era have done. The fact that Manning and Ada had 11 children when they relocated to Debert in 1913, helped to get this family off to a large number. By the time their life at the cottage had started, all their children, except their youngest daughter, Pauline, had married and had families of their own.

As the grand children grew and had families of their own, by 1970, the Geldarts had quite a clan. Although many of the first generation lived in the Debert area, by the second generation, more and more Geldarts were moving out of the area. Once the Second World War was over, family members relocated for employment and sometimes marriage took them to other places.

In 1970, it was decided that it was time for the Geldarts' extended family to reunite. The first reunion was organized for the summer of that year.

The reunion was held at the Coal Mine site on Debert Mountain. This had been the location where Manning, Ada and their 11 children spent their first winter. Family members came together from all over and had a chance to meet other family members they had not seen for years and members they had not yet met. With 40 some first cousins in the second generation, plus their families, the day was a great success.

In the years following, Reunions were held every 2 years at the cottage, the first Sunday of August. This date followed the Debert Field Day and worked well as many people could visit with friends as well as family when they returned home.

At the Reunion, each of Manning's children are given a color code and each direct descendent of that family member wears a name tag of that color to signify which Geldart clan they belonged to.

Reunions, of course, always involve eating and the Geldart Reunions were no different. The supper meal was pot luck and amazingly, there always seemed to be the right amount of different dishes. Long tables were set up outside if the weather permitted and the desserts were set up inside the cottage.

Of course, some reunion years were more significant than others and the day always reflected the life of this family. After the cottage was moved back for the second time in 1976, the reunion was very special. Sid and Lil's daughter, Donna Lou, had married a musician, Brian Johnston. Brian and his band wrote and performed a special song for the occasion. Different years, different people performed depending on their talents. Manning's great-great granddaughter, Carling, Todd and Donna Gratto's daughter, sometimes played the violin. Another of Manning's great- great granddaughter, Alicia, Shelley Geldart's daughter, also played violin. Manning's grandson, Roland, had a daughter-in-law who wrote poetry and sometimes Ann would read a poem that she had written about Portapique or the family.

One year, Lillian had a contest to see who in the family looked the most like her father Manning. Beulah's son Ronnie Graham was the winner. Ronnie also read a special writing that he had done about the cottage.

Many milestone birthdays and anniversaries were celebrated on Reunion Day over the years. Many deaths that had occurred over the months between the reunions were recalled and their absences were deeply felt. In August of 2002, the day before the Reunion, Mae's husband, Laurie, died at the age of 82 after a short battle with liver cancer. Laurie had been very fond of the cottage and over the years had helped a great deal with the upkeep of the cottage. He certainly had fond memories of his time spent on this shore. After the death of Mae in 1986, his days of carpentry were pretty well over, but, he still helped with odd jobs at the "Seashell McCully." The cottage now seemed to have even sweeter memories for Laurie and his daughters. I think they could still sense Mae's presence around them there.

During the 1980's and 1990's, it fell to Lillian and her daughters, Irene and Priscilla to organize these Reunions. Although the Reunions still continue, I am afraid after the next tide, I will not be present for the next one and this deeply saddens me.

20. WHO'DA THUNK IT

One of the greatest changes, here in Portapique, in the last forty years has been the increase in population of cottages in Shady Nook. There are now cottages all along the south facing shore line. Some have changed names with the times and new ones have been added. Manning's grandson, Roland Geldart, had built the cottage to the left of the Seashell McCully but, it has since been sold to an unrelated family. Manning's granddaughter, Sibby , and his great granddaughter, Sylvie, now share the Driftwood Saga that Currie built. Manning's grandson, Art Fisher, Sibby's brother, also has a beautiful cottage a few doors down. There are many cottages now along the road leading in and they have a great view of the marsh and the Portapique River. Priscilla and Eric have a sweet family cottage there called the "Clam Shell".

The names and faces I see visiting these days are different from those of 70 years ago but their expressions are the same as the Geldarts before them. I still see the pleasure on their faces when they visit. The cottage has changed both inside and out when it comes to paint colors but the design has remained basically the same. Even much of the inside furniture has remained the same with a few additions. Lillian's son, Everett, has built several interesting pieces of furniture that he has donated to the cottage. Despite having had Parkinson's disease for many years, his woodworking and creative skills are outstanding and his creations are quite unique. My favorite item that he made for the cottage is the tide clock which hangs on the verandah. I see people checking it each day to make sure the tide is doing what the clocks says. The most constant clock checker is Lorian, Mae's daughter. To Lorian, everything has a proper time and sequence and she makes damn sure the tide holds true to that rule.

Families have increased their vacation seasons here as well, starting in early May and ending after Thanksgiving. This longer season has certainly increased my enjoyment of this family.

The light house and dance hall are long gone as are the old fishing boats. However, in 2006, I had the best surprise. I was sure I was seeing things when "The Juanita", one of the original shad fishing boats from the early 1900's sailed past on a high tide. She had been restored by Russell Cook who lives in Highland Village. She had been the boat, Demi Moore, the actress, sailed aboard when they did the remake of the movie "The Scarlet Letter". Not that I want to dwell on the past, but, my heart was filled with joy at the site of her sailing by and my mind travelled back to the days of yore for several hours. These days, the only boats I see are motor boats and Seadoos, but they just never look as comfortable on the tide as the old sailing ships.

I have seen many things in these waters; mink whales, porpoises, seals, and of course, lots of debris floating that the tide has fetched from other shores, including several notes in a bottle. I have even seen deer swimming on the outgoing tide and a moose drinking salt water when it was driven mad by insects.

Not all strange occurrences happen in the water. In the 1950's, Bobby Graham, Manning's grandson who had been the telegraph boy in Debert, landed his small plane on the beach when the tide was low. Bobby had become a pilot and flying was his passion. He confessed to the family, his landing on Portapique Beach was one of his most difficult, due to the slope of the beach. He later went on to became the private pilot for Prime Minister, Pierre

Trudeau, during the 1970's. In 2006, another small plane landed on the beach. This plane came from Stanley Airport, across the bay and it was flown by a lady pilot. I don't know who was more excited, Priscilla's grandchildren or myself. Just goes to show if you watch long enough you might see things you would never expect in places you would never expect to see them.

Another rare occurrence was the day the tide never went out and unfortunately, I missed that event. An older gentleman had been visiting Manning at the cottage. When Manning and some others left to do the errands, the man was told to dig the clams for supper, at low tide. Upon their return, the tide was full again but there were no clams. When Manning questioned his friend, the gentleman replied "It was a funny thing, but, the tide didn't go out that day"- so sorry I missed that one.

Most amazing off all, when I look out over this Cobequid Bay, is the knowledge that the Seashell McCully, which has been moved back twice, was originally built about 300 feet out into the water. The swelling seas have claimed so much land.

Lorian Randall

21. LAST TIDE

Time and tides on this shore have kept a constant rhythm.

The times have greatly changed in the outside world and some of those changes are evident here on the shores of the Bay of Fundy.

The icecaps have been slowly melting, causing oceans everywhere to swell and shorelines to recede. Although these receding shore lines have been a constant occurrence, in recent years, the increased speed of the melting caps, from global warming, has brought increased speed in the erosion of the shorelines. This erosion has definitely left its mark on Portapique shores, especially where the Geldart cottage is located. The cottage itself is safe for now, with a good 70 feet left in front. But, my demise has been looming for these past few years.

The increased erosion is certainly my greatest enemy, but, today, the wind and sea have turned nasty and my life can now probably be measured in hours. The force of these tides is nothing short of miraculous and they are as constant as the Sun and Moon that pull them up and down this beach. Men on this shore have tried from time to time to minimize that force. Sid, Lillian's husband, built a breakwater in the 1970's but, the January storm of 1976 removed the last pieces that stood on the beach. Others have brought in large pieces of concrete and boulders to shore up the banks. This method has worked to delay the erosion slightly. However, I have seen these tides move those boulders and pull them down the beach. Since the

Geldart's, Seashell McCully and Currie's, Driftwood Saga, have no rock, their banks' erosion has been even greater. Water, like anything else, will always take the path of least resistance. All the man made efforts are only temporary measures and I know in the end, the sea will have its way, at least in my situation.

When I think of this constant rhythm of the tides, I remember Manning Geldart telling his granddaughter Mae, who was supposed to be digging clams, "Time and tide wait for no man and damn few women". These are wise words and we all need to take heed.

The force of the wind off this bay can be amazing and it has also worked against me. It has bent my upper branches so much, I stand at an angle and my roots are now exposed.

It is October of 2008. A severe storm is again beating its way up this coast and because the moon is in its new phase, the tides are at their highest. I am perched on the edge of the bank at a 45 degree angle and I know this tide may be my last. Today, after a hundred years, my life in Portapique may cease.

It almost seems ironic that my greatest joy in life had been being a part of this Geldart legacy and yet, it is man himself who is partly to blame for this increase in erosion. It is not that I hope a miracle will save me from this storm. Nor do I wish I could stop the flow of the wind and sea; that will have to be the challenge of future generations. It's just that my life on this shore has been a good one and saying goodbye is never easy. It will pain me to let go, knowing other trees and even the cottage may lose this battle of erosion.

Whatever happens as a result of this next tide, I know I have been truly blessed by having known this cottage and the families who have graced her doors. For that, I give thanks!

EPILOGUE

October 2009

Another spring and fall have returned to Portapique and the white, spruce tree is still lying on the beach at Shady Nook. The roots still cling to the strands of soil that were pulled out with it. The high tides have washed over it and the salt and the sun have bleached its natural color. It has become like the broken seashells that lay at the tide's edge, as it ebbs out again.

The cottage is now 76 years old and certainly shows her age. I sometimes think the only things that are holding her together are the layers of paint and wallpaper. The verandah is still the best spot for having your coffee or tea (or whatever) in the morning and evening as you watch the sun rise or set. The boards are worn thin and definitely need replacing. The wooden foundation is starting to give way and the floor needs to be leveled.

The kitchen is no longer separated from the main room. An L shaped counter with wooden benches replaced the long wooden table of the 1930's. One of the old metal chairs that Donis had brought so long ago from New Brunswick still sits in the kitchen. It is the chair reserved for the chef of the day, who gets to manage the meal from behind the counter.

The old, tin cupboard, plus many coats of paint, is now used in the main room to hold games, puzzles and extra dishes for the reunions. The framed poem and picture that Greta brought in the 1950's still graces the wall, as does the only kerosene lamp that is left from the railroad.

Almost every inch of wall space is home to multiple photos, paintings and souvenirs; each one reflecting a time or talent in a family member's life. The view from all the large windows is priceless. The new mixes with the old, and, the inside mixes with the outside to create a warm, welcoming retreat.

Sometimes, we too, like the tree and the seashells, are broken down or adrift, but, like the cottage, our true value lies, just in our presence.

For me, this broken tree is a constant reminder of the effect of our warming planet and what we may lose in our near future because of its power.

Although this tree will no longer see the Geldart family as it continues to summer at the Seashell McCully, its life and death on this shore, has not gone unnoticed. Its silhouette against an evening sky will be missed by those Geldarts who hear the call of the sea and must return each summer.

"....It's always ourselves we find in the sea." E.E. Cummings.

Account of "The Seashell McCully" - by Lorian (Matthews) Randall. A Fourth generation Geldart.

UPDATES

In the summer of 2010, one of Manning 's great-great grandchildren proposed to his future wife on the beach at Portapique. In July of that year, Matthew MacLachlan, the oldest son of Lynn (Matthews) and Robin, was home from Calgary where he worked and was visiting his parents in Great Village, having arrived home with his girlfriend, Erin Power, also from Calgary. They jogged from Great Village to Portapique to visit Lorian (Matthews) and Rick who were on vacation at the cottage. It was an extremely hot day and upon arriving at the cottage they decided to go for a swim to cool down even though the outgoing tide was almost on the flats. When they returned from their dip in the bay, Erin's eyes had tears as she showed off her new engagement ring. Matthew had proposed on the beach!

The cottage lost two Geldart grandsons in December of 2010. Ron Graham, who had been ill for some time passed away on December 2nd at home, in Bedford. On December 20th, John (Jack) Geldart was killed near his home in Lincoln, New Brunswick. Jack and his wife Jean were always faithful in returning to Portapique every fall and had a chance to visit with family in this area. Ron had usually gone to the cottage with Jack to keep him company when they were boys and the two had remained great friends well into their eighties. Their loss will be greatly felt at the cottage, especially at the next Geldart Reunion.

The shoreline is still eroding at a scary pace. We are down to about 60 feet in the fall of 2010.

A high tide in late November took the old Barbeque Pit

that Currie had built in the 1960's.

We have also lost two more trees since the tree in this story left this shore.

Mannings great grandson – Kurt Gratto and his family are building their own cottage at Shady Nook and have a beautiful view of the marsh.

APPENDIX

The following is a list of the children and grand children of Manning and Ada Geldart.

Donis (1889-1963) - Greta McCully (sister to William McCully):
-John (Jack)

Percy (1892-1972) - Margaret Douglas:
-Madeline
-Donis

Helen (1886-1954) - William McCully (brother to Greta McCully):
-Gwendalyn
-Eugene
-Carl
-Charlotte
-Helen (Joyce)
-Leah (Hope)

Beulah (1898-1976) - John (Jack) Graham:
-James (Jimmy)
-Glenn
-Ronald (Ronnie)
-Robert (Bobby)
-Williams (Billy)
-Ellen

Walter (1900-1986) - Frances Matthews (1st cousin to Laurie Matthews):
-Roland
-Delma
-Vaughan
-Eric
-Delena
-Joan

Altie (1902-1977) - Helen Fisher (sister to Fraser Fisher):
-Rettie
-Frank
-Howard
-Marilyn

Adelaide (1903-1948) - Fraser Fisher (brother to Helen Fisher):
-Elizabeth (Betty)
-Albert
-Donald
-Gerald (Jerry)
-Shirley
-Sylvia (Sibby)
-Arthur (Art)
-Paul
-Carol
-John

Currie (1906-1981) - Margaret Edwards:
-Jeanne - daughter Slyvie

Lula (1907-1988) - Alden Dewar:
-Lois
-Joan
-Doreen
-David

Lillian (1910-2003) - Sidney McCully:
-Leonard (Len)
-Alice (Mae) - Laurie Matthews - daughters Lorian and Lynn
-Irene - Aubrey Gratto - sons Todd, Kurt and Noel
-Phillip (Manning)
-Linda
-Priscilla - Eric Jennings
-Everett
-Michael
-Donna - Brian Johnston

Made in the USA
Charleston, SC
18 June 2015